RAND

Perspectives on Theater Air Campaign Planning

David E. Thaler, David A. Shlapak

Prepared for the United States Air Force

Project AIR FORCE

PREFACE

This report describes the findings of an independent, exploratory look at how the United States Air Force plans and executes air campaigns. It is not a primer on campaign planning, although it does discuss—briefly—some of the processes involved. Nor does it pretend to exhaustively examine all aspects of this complex topic. Instead, the work is intended to identify and illuminate some issues of concern to planners, serving as fodder for future, more detailed analyses.

The work was sponsored by the Director of Plans, U.S. Air Force.

This document should benefit the Air Force in several ways. First, it surveys the state of the art in planning and executing air operations two to three years after the Gulf War. During this time, planners have internalized some of the lessons learned from the desert conflict and have begun to apply them to their own unique situations. It seems an opportune time to take stock of the progress made.

Second, the report is intended as a vehicle in which planners in the air component commands, the Air Staff, and elsewhere can express their current concerns to a broad audience of analysts and decision-makers. We by no means purport to be "speaking for" planners—the observations here are naturally colored by our own perspective, and the conclusions are entirely our own. However, we do believe that the document provides a unique opportunity for expressing common concerns.

Third, we hope that the various staffs involved in planning and conducting air campaigns will use the information contained in these pages to learn from each other. We sometimes found that individual organizations solved problems by developing workarounds of which other organizations with similar problems were unaware.

Finally, the document recommends areas of focus for future research—its key purpose. As such, we hope that it will guide future efforts within the Air Force and in RAND's Project AIR FORCE.

PROJECT AIR FORCE

Project AIR FORCE, a division of RAND, is the Air Force federally funded research and development center (FFRDC) for studies and analyses. It provides the Air Force with independent analyses of policy alternatives affecting the development, employment, combat readiness, and support of current and future aerospace forces. Research is being performed in three programs: Strategy, Doctrine, and Force Structure; Force Modernization and Employment; and Resource Management and System Acquisition.

Project AIR FORCE is operated under Contract F49620-91-C-0003 between the Air Force and RAND.

CONTENTS

FIGURES

SUMMARY

This document reports on independent, exploratory research on air campaign planning and execution. Its purpose is threefold: (1) to provide observations on the current processes and capabilities for planning and executing air operations in theater conflicts; (2) to identify key issues associated with those processes; and (3) to recommend analytic concentrations for future research. Our primary focus is at the broadest level of campaign planning and execution—activities flowing from the definition of campaign and operational objectives down to the allocation, apportionment, and tasking of forces.

Our work involved interviews with a variety of USAF organizations.[1] We did not explicitly interview personnel in other service or joint organizations; hence, we are plainly offering a USAF perspective on campaign planning. However, most of the USAF entities we talked with are intimately involved in joint and combined operations and planning, and Air Force doctrine and perceptions will likely play an important role in shaping future air campaigns. Therefore, we believe that this report has relevance beyond the confines of the USAF planning community.

[1]All our discussions were on a nonattribution basis; therefore, we do not identify by name the source of any comments. In fact, we identify organizational remarks or perspectives only when we believe that doing so is integral to establishing or understanding a key point.

OBSERVATIONS ON U.S. AIR CAMPAIGN PLANNING AND EXECUTION

Defining and Articulating Objectives

Defining clear and coherent objectives is perhaps the most critical step in crafting an effective air campaign plan. In this regard, the chain between the national command authority (NCA), the theater commander-in-chief (CINC), and his components can break in at least three places: (1) The NCA may not articulate national objectives clearly enough for the CINC to develop well-defined and executable campaign plans; (2) the CINC's guidance may be unclear to one or more of his component commanders; and (3) the components may be unable—or unwilling—to harmonize their activities to achieve the most effective application of their joint combat power in service of the CINC's intent. In our conversations with planners at the different commands, we discovered examples of all three disconnects.

Additionally, we found that measures of effectiveness for achieving objectives often lack any specific operational basis and are sometimes drawn from the individual experience of a senior commander ("Deliver *X* weapons of *Y* type on each target in category *Z*.") Although such measures of effectiveness may be perfectly valid—based as they are on a commander's expert judgment—they provide no clear basis for relating the outcome of a given attack to any effect on enemy combat capability.

Integrating Intelligence

Establishing and maintaining an adequate intelligence support infrastructure are prerequisites to selecting appropriate strategies and defining and tracking attendant measures of effectiveness and outcome. Although we did not set out to evaluate intelligence support at the national and theater levels, we did take a few opportunities to understand the level of integration between the intelligence communities and planners at the working level. It is in this area that we found the greatest divergence in perspectives between our Pentagon contacts and those in the field.

Officers on the Air Staff repeatedly emphasized the difficulties they have experienced getting what they perceived as adequate support from the intelligence community. In the field, integration appears better, if still not seamless. Intelligence staff work side-by-side with the planners, learning their needs and at times growing adaptive enough to anticipate them.

At each site we visited we heard concerns expressed that, as one individual put it, the "feedback loop" between planning, execution, and assessment "is broken." That is, accurate information about what has been attacked, and to what effect, does not percolate rapidly back to either the target-nominators or the planners. This can lead to unnecessary reattacks, or neglect of a still-functioning target. Additionally, friction can arise between target-nominators and planners when the former cannot tell from assessment reports that their input is being acted upon. Intelligence personnel we spoke with shared with the planners the desire to develop a more functionally oriented approach to damage assessment; indeed, they recognize this as one of the foremost challenges they face as a community.

Assessing the Responsiveness of the Execution Cycle

The process during a campaign of setting priorities, developing a time-phased plan, and building and disseminating an air tasking order (ATO) is sometimes referred to as the planning, decision, and execution (PDE) cycle. In and of itself, a PDE cycle can run from about 36 to 48 hours, depending upon whether certain high-level CINC decisionmaking is included in it.

Many observers believe that the PDE cycle should be accelerated. We believe that *flexibility* is really at the heart of the responsiveness issue, not the number of clock-hours it takes to produce and distribute an ATO. Flexibility can be enhanced both by increasing the speed of ATO generation and by restructuring the planning process to have more built-in adaptability and agility. This is not to argue that current initiatives to speed up the planning cycle should be abandoned. Rather, the PDE cycle should be placed in an appropriate context, one that focuses on efficiently accomplishing operational tasks and objectives.

Automating Support to Planning

We found universal enthusiasm for the Contingency TACS (Theater Air Control System) Automated Planning System (CTAPS) among the groups we visited. CTAPS is designed to facilitate intelligence availability, ATO construction and dissemination, and airspace management, among other things. All information will be available to all users on an as-authorized basis (e.g., F-15E planners may be able to view, but not change, tanker-related information). If properly developed, systems like CTAPS can integrate a variety of heretofore disparate functions into an architecture that provides planners with a degree of support previously unknown.

An initiative that shows promise in the area of objective definition and articulation is the air campaign planning tool (ACPT) being developed in the Air Staff. The ACPT is a minicomputer-based tool that helps the user link objectives to one another and also highlights areas where connections, or the objectives themselves, are missing or unclear. The user also sets priorities among objectives and draws on a large database to identify appropriate targets for achieving specific goals. The end result is an overall prioritized target list that is linked up through the hierarchy to overall campaign goals.

Understanding Organizational Perspectives

We found the four principal organizations we visited—Ninth Air Force, USAFE (United States Air Forces in Europe), Seventh Air Force, and HQ PACAF (Pacific Air Forces)—to be in substantive agreement on most of the issues we discussed: intelligence support, prospects for automation, and so forth. However, we did find some differences in tone and approach that are worth noting.

One unique emphasis we found at Ninth Air Force was on the power of personalities to affect a campaign. This perspective appears strongly informed by the organization's experience in *Desert Shield* and *Desert Storm*, when new procedures, organizational arrangements, and personnel were overlaid from the top down on existing structures. The lesson here seemed to be that the best laid plans, structures, and associated training can be easily overturned by commanders who, rightly or wrongly, have strongly held convictions about what is needed to support them. Planners everywhere must be prepared for such situations.

Seventh Air Force's perspective was shaped by the history and geography that place the command in the center of a well-developed theater, with an up-and-running coalition and a viable threat literally minutes away. Of the three commands, it most strongly emphasized the importance of joint and combined operations. Korea was also where we encountered for the first and only time worries about rear-area security and air base defense. Deep concerns were expressed about the danger posed by North Korean special forces (SF).

We are concerned that current deployment plans may be somewhat inflexible and therefore vulnerable to derailment by the kinds of disruptions in the allied rear area that we have described. This danger could be especially severe in the event of a short-warning attack, already a very stressful scenario. Exercises should take these possibilities into account so that any deficiencies can be identified and addressed before a crisis flares.

With the enormous changes in the European security context, USAFE finds its setting and responsibilities in flux. For forty-plus years, U.S. forces in Europe planned against the Warsaw Pact; their circumstances then were in many ways analogous to those prevailing today in Korea—imminent threat, in-place coalition, well-understood planning concepts and objectives, and a "fight-in-place" mind-set. Today, their situation bears more than a passing resemblance to that of Ninth Air Force: USAFE is now the air component of an expeditionary force that must prepare to fight anywhere, anytime, with a pickup team of allies. This is a fundamental, and wrenching, kind of change to which the organization is still adapting.

RECOMMENDATIONS FOR FUTURE ANALYSIS

We identify four areas that are amenable to in-depth analysis to improve air campaign planning and execution.

- Define, prioritize, and establish the political relevance of military objectives for a wide range of scenarios.

A crucial challenge is to provide remedies for situations where national- and theater-level objectives are not well defined or where cause-effect relationships between military options and desired political results are unclear. This points to the need to build a menu of potential campaign and operational objectives and attendant measures of outcome in various scenarios, to gain insights into appropriate priorities among these objectives, and to link the achievement of these objectives to political aims.

- Develop new concepts for functional assessment of the results of battle.

In many ways, the treatment of objectives is closely linked with efforts to meet the second challenge—enhancing intelligence support to commanders, planners, and operators. Many planners we encountered spoke of the need for a greater focus on *functional* as opposed to *physical* results of battle. Commanders and planners need to know the effect of their actions on enemy capabilities, not merely how many items of enemy equipment are "confirmed kills." They require information about the status of a target system, how the status is changing, and how this relates to attainment of the commander's objectives. Only then can he adjust his strategy in the most effective way. Fundamentally, then, intelligence analysis should focus on assessing the output of a targeted entity or system, not its physical integrity.

- Develop new concepts for improving the flexibility of the planning, decision, and execution process by both speeding up the planning cycle and making the plans more inherently adaptive.

We believe that efforts focused simply on "speeding up" the PDE are inadequate to meet future planning challenges. The goal must not be just to prepare the ATO more quickly, but to develop a planning process that (1) provides timely enough outputs to allow those charged with generating sorties to do so, (2) allows the CINC and component commander to oversee the overall campaign architecture, (3) provides greater visibility into both the planning process and execution outcomes to the other components, and (4) is sufficiently adaptive to permit appropriate responses to changing circumstances in the battle space. In addition to completing current initiatives aimed at building plans more rapidly, we therefore suggest implementing a planning process that decentralizes the assignment of weapon platforms to specific targets and placing greater reliance on real-time control by mission control elements on the ground and in the air.

- Identify and refine options for full utilization of air operations groups.

Each command we visited has organized or is organizing an *air operations group* (AOG), a kind of standing mini-JFACC (Joint Forces Air Component Commander) staff. The AOG can serve at least four important functions in preparing the ground for more effective air campaign planning. First, it can serve as a vehicle for resolving disputes and disconnects between communities (intelligence and operations) and components (air and ground) before an actual crisis or conflict brings the issues to the fore in a more dramatic and costly fashion. Second, the AOG might be given a long-range planning role, providing the command with a cell that is looking ahead, beyond day-to-day issues. Third, building on both of the above, the AOG can fill an obvious training role for JFACC staff. Finally, the group could be used as a laboratory for testing out new planning and execution concepts.

ACKNOWLEDGMENTS

The authors are indebted to a number of people who devoted precious time and energy to them during the course of this study. Colonel Joe Bonner at Ninth Air Force, Colonel Bob Hylton and Major Tom Twohig at USAFE, Colonel Bob Gregory and Lieutenant Colonel Tom Braund at Seventh Air Force, and Major Jim Miyamoto and Captain Joe Sousa at HQ PACAF all went to considerable lengths to ensure that the authors' visits to these organizations were productive and comfortable. Many others on the Air Staff, at Air University, and elsewhere were kind enough to offer their insights as well. Finally, the authors appreciate the hours RAND colleagues Glenn Kent and Edward Warner spent offering their thoughts on campaign planning.

Alan Vick and Myron Hura of RAND provided careful and insightful reviews of the draft version of this report.

The authors appreciate the support provided by the study's Air Force facilitators, Lieutenant Colonel Mike Nelson and Lieutenant Colonel Bob Bivins (AF/XOOC). Finally, we are indebted to Glenn Buchan of RAND and Colonel Charles Miller (AF/XOXP) for their encouragement and countenance.

ACRONYMS

ACC	Air Component Commander
ACPT	Air Campaign Planning Tool
AFB	Air Force Base
AOC	Air Operations Center
AOG	Air Operations Group
AOR	Area of Responsibility
ASOC	Air Support Operations Center
ATO	Air Tasking Order
AWACS	Airborne Warning and Control System
BDA	Bomb Damage Assessment
CENTAF	U.S. Central Command Air Forces
CINC	Commander-in-Chief
CRC	Control and Reporting Center
CTAPS	Contingency Automated Planning System
ITO	Integrated Tasking Order
JFACC	Joint Force Air Component Commander
JFC	Joint Forces Commander
JSTARS	Joint Surveillance and Targeting System
MAP	Master Attack Plan
MCE	Mission Control Element
MOE	Measure of Effectiveness
NCA	National Command Authority
PACAF	Pacific Air Forces
PDE	Planning, Decision, and Execution
SAM	Surface-to-Air Missile
SF	Special Forces
TACS	Theater Air Control System
TLAM	Tomahawk Land-Attack Missile
USAF	United States Air Force
USAFE	United States Air Forces in Europe

Chapter One

INTRODUCTION

Operation *Desert Storm* was a watershed event for the application of air power in pursuit of U.S. national security and military objectives. Observers have commented extensively on the superior training and professionalism of U.S. military personnel, the innovative concepts for accomplishing operational tasks under demanding conditions, and the technological advances manifested in stealth and precision strike capabilities. The war in the desert also highlighted a most critical element of the effective employment of air power—how the United States plans and manages air operations to attain U.S. goals. The Gulf War breathed new life into efforts to ensure that the concepts and procedures for theater battle management of air operations would take full advantage of the capabilities that U.S. personnel and equipment bring to a campaign.

At the same time, *Desert Storm*, like every campaign, was planned for and executed under unique circumstances. A combination of factors provided favorable conditions in which to plan and conduct military operations. Objectives were clearly articulated and actionable at all levels—from the President on down. The United States had nearly six months to develop and refine campaign plans, arrange organizational structures and relationships, and train its personnel for the tasks at hand. The time and force structure available allowed systematic deployment of nearly all of the combat and support forces that planners indicated were needed to do the job. Finally, the United States was faced with an adversary who seemed bent on digging into the sand and ceding the initiative.

Obviously, the likelihood is low that these factors will again array themselves so propitiously. Potential U.S. adversaries certainly have learned valuable lessons from Saddam Hussein's blunders, as well as from U.S. successes. Moreover, current and future drawdowns in U.S. force structure will in many situations provide planners and operators with fewer forces to be employed against these smarter foes. As a result, the United States will place a premium on finding ways of deploying and employing forces more effectively and efficiently. Air power will in many cases be pivotal to the success of U.S. military operations; hence our focus on how the United States plans and executes air operations in support of theater campaigns.[1]

This study reports on exploratory research on air campaign planning and execution. Its purpose is threefold: (1) provide observations on the current processes and capabilities for planning and executing air operations in theater conflicts; (2) identify key issues associated with those processes; and (3) recommend analytic concentrations for future research. Our primary focus is at the broadest level of campaign planning and execution—activities flowing from the definition of campaign and operational objectives down to the allocation, apportionment, and tasking of forces.

We provide an overview of campaign planning and execution in Chapter Two. This overview establishes common terms of reference associated with the relevant processes. In Chapter Three, we offer observations on the current processes, procedures, and capabilities used by planners. We divide these observations into five subject areas: (1) defining and articulating objectives, (2) integrating intelligence, (3) assessing the responsiveness of the execution cycle, (4) automating support to planning, and (5) understanding organizational perspectives. We offer our insights into some of these areas in Chapter Four, and within this context, recommend areas of emphasis for future research. Concluding remarks appear in Chapter Five.

Our work involved interviews with a variety of United States Air Force (USAF) organizations, including Ninth U.S. Air Force (Shaw Air Force Base (AFB), SC), United States Air Forces in Europe (USAFE,

[1]Although our focus is on large-scale air operations, we believe the bulk of what follows is applicable across a wide range of combat air operations, including smaller ones.

Ramstein AFB, Germany), Seventh Air Force (Osan AFB, Republic of Korea), Pacific Air Forces (PACAF, Hickam AFB, HI), and various offices in the Air Staff.[2] We did not explicitly interview personnel in other service or joint organizations; hence, we are plainly offering a USAF perspective on campaign planning. However, most of the USAF entities we talked with are intimately involved in joint and combined operations and planning, and Air Force doctrine and perceptions will likely play an important role in shaping future air campaigns. Therefore, we believe that this report has relevance beyond the confines of the USAF planning community.

---

[2]All our discussions were on a nonattribution basis; therefore, we do not identify by name the source of any comments. In fact, we identify organizational remarks or perspectives only when we believe that doing so is integral to establishing or understanding a key point.

Chapter Two

AN OVERVIEW OF AIR CAMPAIGN PLANNING AND EXECUTION

In this chapter, we offer an overview of air campaign planning and execution to help establish a reference point from which to build our treatment of the issues. In a way, this overview describes our initial, somewhat idealized image of how planning is, or should be, conducted and represents the mindset we had while pursuing this research.

THE BACKDROP FOR PLANNING: A HIERARCHY OF OBJECTIVES

Forces are deployed and employed to achieve objectives. These objectives constitute the backdrop against which campaigns are planned and executed. One can portray a hierarchy of objectives that links broad national security and military objectives to specific operational tasks at the tactical engagement level.[1] The hierarchy flows as follows:

1. The President formulates *national security objectives* toward which the United States applies its national power to secure fundamental national goals (e.g., providing for the common defense) in the face of threats to those goals and opportunities for furthering them.

[1]This hierarchy of objectives is described in detail in David E. Thaler, *Strategies to Tasks: A Framework for Linking Means and Ends*, Santa Monica, Calif.: RAND, MR-300-AF, 1993.

2. The Secretary of Defense (SecDef) and Chairman of the Joint Chiefs of Staff (CJCS) define *national military objectives* that guide the application of U.S. military power in various regions to support national security objectives.
3. The unified and specified combatant commanders define *campaign objectives* that guide the prosecution of campaigns for attaining national military objectives in their areas of responsibility (AORs).
4. The combatant and component commanders define *operational objectives* that guide the posturing and employment of forces to gain the campaign objectives.
5. Elements of the component staffs define *operational tasks* that force elements are assigned to accomplish to achieve the operational objectives.

Strategies link the levels of the hierarchy; an objective is attained through the implementation of a strategy. A strategy for achieving a *supported* objective is in part a statement of *supporting* objectives. For example, the President's strategy for attaining the objective of freeing Kuwait in 1991 was to evict Iraqi forces from Kuwait, isolate the Iraqi regime diplomatically, and levy economic sanctions on Iraq. Although evicting Iraqi forces was a component of the national *strategy,* for the combatant commander this was an *objective.* In other words, objectives cascade—what is a strategy at one level becomes an objective at the next lower level.

Obviously, a strategy is not merely a statement of objectives—it also defines *the weight of effort to be applied over time* among objectives at one level to attain the higher-level objective *in a given situation.* Weight of effort refers to the relative priority accorded an objective and the level of resources (forces) allocated toward achieving it. Weight of effort *over time* expresses the notion that these priorities and attendant resource allocations may change according to the situation. For example, a commander may stress gaining air superiority in the beginning of a campaign, and later shift this emphasis to defeating ground forces due to the results of battle. Strategies, then, serve as the link between levels of objectives in the hierarchy and provide the context in which the objectives are achieved.

CAMPAIGN PLANNING AND EXECUTION: A HIERARCHY OF PROCESSES

A set of processes for designing and taking actions in specific scenarios accompanies each level of the hierarchy of objectives and strategies. Therefore, we refer to a hierarchy of processes for campaign planning and execution at the strategic/campaign level, the operational level, and the engagement level; some processes cut across all levels.

Figure 2.1 graphically depicts our image of these processes. The figure represents both peacetime and wartime activities. In describing these activities, we may omit a number of details; our purpose here is to build a general baseline for our discussions in Chapters Three and Four.

Processes at the Strategic/Campaign Level

In peacetime, the Secretary of Defense and the Chairman of the Joint Chiefs of Staff direct unified and specified combatant commanders (also referred to as commanders-in-chief, or CINCs) to plan in support of national military objectives in the commanders' AORs. This tasking comes in the form of broad policy guidance as found in the *National Military Strategy* and the Defense Planning Guidance, wherein the CINCs are directed to plan for certain scenarios and are given broad assumptions about the forces expected to be available. When the National Command Authority (NCA) chooses to employ U.S. military power, guidance to the responsible commander defines national military objectives specific to the conflict, criteria for attaining these objectives, and political and moral constraints, among other things.

Accordingly, a combatant commander prepares plans for conducting campaigns in his assigned AOR. The development of these campaign plans involves planners and intelligence specialists on his staff as well as the air, land, and sea components under his command. The CINC defines campaign objectives and the weight of effort among them to attain the national military objectives set forth in the NCA guidance. He also describes his vision of how forces assigned to him will be apportioned across these campaign objectives. The CINC

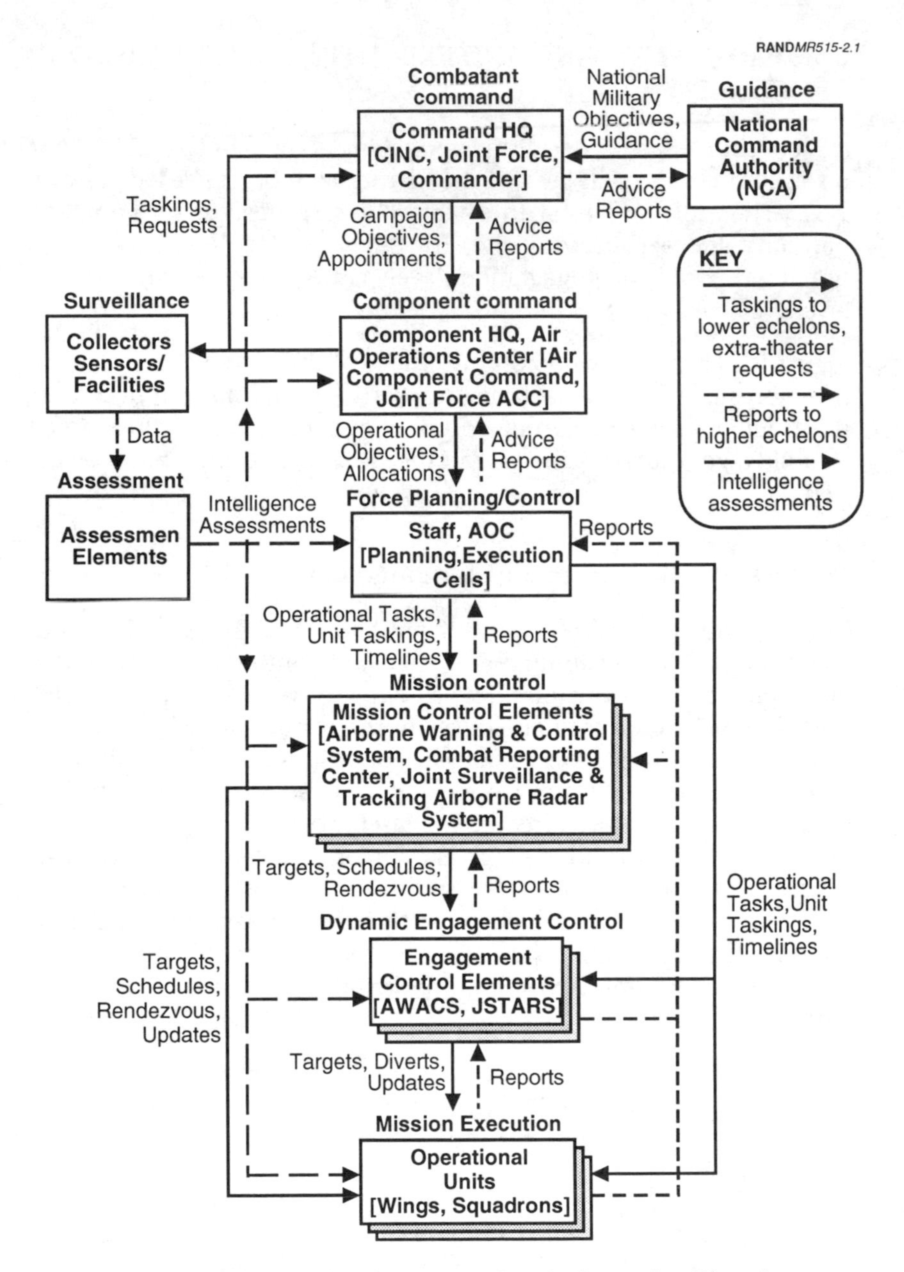

Figure 2.1—Hierarchy of Processes for Air Campaign Planning and Execution

selects the scheme of maneuver of ground forces, apportions and allocates air forces among operational objectives and geographic areas, and directs naval and amphibious operations as appropriate. Thus, the campaign plans contain the combatant commander's operational strategy for employing forces in his AOR.

The CINC must also communicate his emphases and assumptions—e.g., forces available, allies, and operational constraints—to lower echelons so that they can conduct more detailed planning. This detailed planning includes identifying the most critical elements of enemy military, political, and economic strength—the enemy's centers of gravity—defining (in coordination with the CINC) operational objectives and tasks that support the combatant commander's campaign objectives and intent, matching force elements to these objectives and tasks and developing a time-phased concept of employment, and drawing up plans for the deployment of these force elements.

In wartime, the CINC—with the aid and advice of his component commanders—will adjust his campaign plan according to the progress and results of friendly operations, changes in policy guidance from above, and the status and actions of enemy forces. This is the essence of theater command and battle management—adjusting the strategy in response to changing circumstances and apportioning forces accordingly. Selection of appropriate strategies and adjustments to those strategies depends on an adequate architecture for gathering information on the enemy, assessing that information, and preparing intelligence assessments.

At times, the component commanders under the CINC may have disagreements over the proper weights of effort among objectives, different targeting priorities, or competing demands for resources. The CINC or his designee arbitrates these disagreements and possesses final authority to decide on the best course of action.

Processes at the Operational Level

The air component commander (ACC) plans the deployment and employment of air power to support the theater CINC's campaign plans, as do his naval and marine counterparts. In wartime, the CINC designates a Joint Force Air Component Commander (JFACC) to ensure that in-theater air assets of all services are deployed and employed effectively according to a single plan—often referred to as an integrated "air campaign."[2]

The JFACC guides the posturing and employment of air power to attain the CINC's campaign objectives. With the aid and advice of his staff in the air operations center (AOC), he defines the weight of effort among operational objectives and allocates air assets accordingly. As reports of battle results, enemy action, and the influx of forces become available, he adjusts his weights and allocations. The AOC staff turns the JFACC's guidance into a more detailed plan that indicates operational tasks to be accomplished, targeting priorities and packages of weapon systems to be used in support, and timelines. In *Desert Storm,* this more detailed plan was called the master attack plan (MAP). The MAP was conceived as a planning tool for translating the JFACC's intent into specific, time-phased operations.

If the MAP is the general "treatment" of a script, the air tasking order (ATO) is the script itself. The ATO guides the aircrews that must go out and execute missions. It is used to

- task attacks on specific targets
- coordinate defense suppression and force protection missions
- schedule tanker tracks and in-flight refuelings
- direct manning of engagement control orbits
- deconflict airspace
- provide call signs and identify friend-or-foe (IFF) squawks.

The planning cell in the AOC prepares and disseminates the ATO to air units. The AOC's execution cell, as its name implies, monitors the execution of the ATO.

---

[2]The JFACC may be from the Air Force, Navy, or Marines, largely depending upon the situation. If Air Force assets constitute the bulk of available air forces, a JFACC from the Air Force might be appropriate. If air forces are primarily sea-based in a given scenario, a Navy or Marine Corps JFACC might be designated. The overall goal in selection of a JFACC is to deploy and employ available air assets under a single, integrated air campaign plan; the appropriateness of the selection is based on this.

The ATO provides the greatest targeting detail with regard to fixed installations. For missions that involve fleeting targets, such as close air support or defense of friendly airspace, the ATO's primary role is generating sorties. Mission control elements (MCEs) manage operations against targets that cannot be specifically tasked against in the ATO, such as mobile force elements. The Control and Reporting Center (CRC) and Air Support Operations Center (ASOC) are examples of ground-based MCEs for air defense and close air support, respectively. E-3 airborne warning and control system (AWACS) aircraft and E-8 Joint Surveillance and Targeting System (JSTARS) aircraft also exercise mission control.

Processes at the Engagement Level

At the engagement level, dynamic engagement control elements such as AWACS and JSTARS may provide real-time assistance to air crews. In defending against enemy air attack, for example, AWACS aircraft help control the air battle by vectoring individual fighters toward incoming enemy aircraft. JSTARS aids ground attack aircraft in locating and identifying moving enemy vehicles. In other cases, engagement control elements may relay updates to air crews from offboard sensors. Such updates could include changes in the weather, intended targets, or enemy defenses. And, if the weather over a primary target deteriorates enough to preclude an effective attack on it, an engagement controller might divert attack aircraft en route to that target to secondary targets.

Finally, force elements prepare for and execute missions assigned to them in the ATO. In preparing for operational tasks involving surface attacks, air crews develop targets, select routes, and decide on tactics. To develop fixed targets, for instance, crews establish appropriate aimpoints and evaluate target status. In selecting routes for ingress and egress, they familiarize themselves with enemy radars and air defenses to avoid and with waypoints to aid in navigation. Crews select appropriate tactics based upon expected weather over the target, the target's characteristics and environment (e.g., geography, point defenses), and the type of weapon to be used against the target. The air crews then fly the mission, often with the aid of the mission control and engagement control elements described above.

Processes Cutting Across All Levels

Three other processes—intelligence gathering and assessment, communications, and training—cut across the three levels identified above.

To conduct air campaign planning and execution in an effective and efficient way, commanders, planners, and operators must be kept informed to the greatest extent possible about all relevant aspects of the enemy's activities and war-supporting infrastructure—from the nature of his political leadership to the status and movement of his forces to the strength of hangar doors on his aircraft shelters. Thus in peacetime, various sources collect such information to provide a basis for judgments regarding possible enemy intentions and the character of the target system of the adversary. Once war begins, data are collected on the performance of the adversary's weapon systems, the tactics he uses, the status of his forces and facilities, and the results of battle.

The raw data are usually transmitted to assessment centers in the continental United States (CONUS) or in theater where intelligence experts evaluate them. Personnel in these centers process, cross-correlate, and evaluate the data, and then present their analysis to commanders and their staffs. These intelligence assessments are combined, in turn, with data from other sources to produce an overall situation assessment that includes estimates of the opponent's capabilities, information regarding the location and activities of his forces, and judgments on likely enemy courses of action. The commander formulates and adjusts his campaign plan on the basis of this information and the availability and capabilities of U.S. and allied forces at his disposal.

These same situation assessments will be used in a crisis to provide indications and warning of an opponent's preparations for attack. Once the war begins, the assessment also includes reports on the recent activities and performance of the adversary's forces and the results of friendly operations. The situation assessment is combined with evaluations of U.S. and allied capabilities in the theater—which change constantly as a result of both enemy and friendly actions—to enable the commander to adjust his plans in an informed way. Planners and operators use detailed information about enemy forces and assets to, among other things, assess which force elements to deploy, assemble appropriate strike packages, and determine the right weapon for the job at hand.

As a practical matter, information on enemy intent and capabilities sometimes is inaccurate or untimely, and rarely is it complete. Intelligence analysts during peace and war endeavor to fill in as much of the puzzle as possible, but key pieces often are missing. It is in situations where hard facts are lacking that the experience and savvy of the intelligence specialist are crucial. He is called upon to make educated guesses on questions ranging from "Is that bridge passable?" to "How is the enemy likely to respond to this course of action?" Without exception, campaigns are planned based on imperfect knowledge of the adversary and his actions, but the intelligence specialist helps fill in the gaps.

The taskings, reports, and information flows represented by the lines in Figure 2.1 are all made possible by a complex communications architecture. As one follows Figure 2.1 from the NCA at the top to the wings and squadrons below, the communications pathways proliferate. Thus, the architecture must be adequate to accommodate conferences between the NCA in Washington and commanders halfway around the world, dissemination of intelligence information to geographically scattered consumers, and two-way communications among commanders, controllers, and engaged forces during the heat of battle.

For all these elements to work and interact properly during war, the aircrews, planners, communicators, intelligence analysts, and commanders must be well trained. Much of the peacetime training focuses on enabling planners, operators, and analysts to hone their skills and work smoothly with one another. The most widely recognized type of training is combat training, where air crews practice their craft. Allowing planners to practice developing and disseminating ATOs, intelligence analysts to train at rapidly formulating situation assessments, and potential Joint Forces Commanders (JFCs) and JFACCs to define objectives and establish proper weights of effort is equally important. Moreover, exercises that allow different players to work together help ensure well choreographed campaigns at all levels. They also help identify areas for improvement during peacetime to reduce the likelihood of surprises during battle.

Chapter Three

OBSERVATIONS

This chapter provides observations and comments on air campaign planning and execution based on our discussions with planners in the Air Staff, at theater component commands, and elsewhere. We formulate our comments within five broad categories:

- How are air campaign objectives defined and articulated? How well are command relationships defined? What measures of effectiveness and outcome are used to track the progress of a campaign?
- How well is intelligence integrated into the planning process?
- What are the prevailing perceptions regarding the responsiveness of the planning process in wartime?
- What benefits are anticipated from increased automation?
- Are there unique organizational perspectives that provide insight into planning requirements?

We will address each in turn.

DEFINING AND ARTICULATING OBJECTIVES

Defining clear and coherent objectives is perhaps the most critical step in crafting an effective air campaign plan. According to Air Force Manual 1-1, *Basic Aerospace Doctrine of the United States Air Force,*

> the key to success lies in an air component commander's ability to achieve *objectives* by orchestrating aerospace roles and missions so they produce a mutually reinforcing effect.[1]

Similarly, the *JFACC Primer* states:

> The essence of the JFACC [Joint Forces Air Component Commander] concept is not simply the designation of a single commander for air. Its broader focus is the development of a Concept of Air Operations to meet the *objectives* set by the JFC. The concept of air operations bridges the gap between assigned strategic objectives and the execution of air operations to accomplish those objectives. *The JFACC is not just in the business of servicing targets.* The concept of air operations is embodied first in the air campaign plan, subsequently in the master attack plan, and finally in the execution ATO.[2]

These statements suggest a process such as that sketched out in Chapter Two. In this discussion, we will focus on the campaign and operational levels of planning and execution.

The chain between the NCA, the CINC, and his components can break in at least three places: (1) the NCA may not articulate national objectives clearly enough for the CINC to develop well-defined and executable campaign plans; (2) the CINC's guidance may be unclear to one or more of his component commanders; and (3) the components may be unable—or unwilling—to harmonize their activities to achieve the most effective application of their joint combat power in service of the CINC's intent.

In our conversations with planners at the different commands we discovered examples of all three disconnects. In Europe, some USAFE staff expressed frustration that objectives for proposed military action in the former Yugoslavia were vague to the point of being virtually useless as a foundation for planning.[3] They observed

---

[1]Department of the Air Force, *Basic Aerospace Doctrine of the United States Air Force*, AFM 1-1, Vol. 1, Washington, D.C., 1992, p. 10, emphasis added.

[2]Headquarters, U.S. Air Force, Deputy Chief of Staff for Plans and Operations, *JFACC Primer*, Washington, D.C., 1992, p. 16, emphasis added.

[3]We note that, despite the lack of clarity from above, those who have developed plans for combat air operations in Bosnia seem to have done a good job defining alternative

that without a well-understood set of goals, the tendency is to hedge by planning against the worst case, which may result in completely inappropriate options. Alternatively, planners may find themselves reduced to merely "servicing targets"—absent real objectives and clear guidance, there may be little else to fall back on.

We also heard how confusion regarding the CINC's intent can be the source of difficulties. Some Ninth Air Force veterans of the Gulf War recalled that Gen. Schwartzkopf would sometimes give directions to one of his component commanders in private, off-line conversations, the content of which was never passed on to other senior officers. So, for example, at one point the CINC instructed the JFACC not to target any more air strikes on Iraqi divisions that were assessed to be below 50 percent strength. However, "not until after the war were corps commanders aware of the CINC's guidance"—prompting many complaints from Army commanders about a lack of air support in their areas.[4] The result was occasional confusion among the components about what each—but most particularly the air forces—had actually been tasked to do.

Finally, exercises in Korea have highlighted ongoing difficulties coordinating Marine tactical aviation and Navy Tomahawk land-attack missiles (TLAM) into the integrated tasking order (ITO—the Seventh Air Force version of the ATO). Marine doctrine strongly emphasizes the use of Marine aviation to support Marine Corps ground operations; only excess sorties are available for joint tasking. Since this number varies depending on circumstances, it is not always easy to integrate Marine air into a joint campaign. As regards TLAM, the Navy apparently has concerns over the security of TLAM targeting information which Seventh Air Force staffers said have proved difficult to overcome. As one observed, the flag officers in the theater seem to

---

military objectives. Colleague Alan Vick makes the point that given the very nature of what are called lesser-regional contingencies and peace operations—multiple subnational actors; an often-nasty stew of ethnic, religious, and historical factors; intricate and opaque political processes; and dispersed and relatively primitive military capabilities—useful military actions of all kinds are difficult to identify. Defining actionable objectives requires at least a minimal understanding of each conflict's unique dynamics; in such cases, fuzzy and obscure guidance may be the norm. Planners will need tools, processes, and mind-sets that allow them to function with reasonable efficacy in this normal, confused, situation.

[4]Col. R.B.H. Lewis, USAF, "JFACC: Problems Associated with Battlefield Preparation in Desert Storm," *Airpower Journal*, Spring 1994, p. 14.

have internalized the importance of joint air operations—and the criticality of unity of command to their proper execution—but "the word" has apparently not yet filtered down to everyone at the working level. All involved expressed hope, if not confidence, that the necessity for jointness would become an inculcated perception across all ranks over time.

The importance of the JFACC surfaced frequently in our discussions. In Korea, the role of the air component commander seems well defined and largely accepted by other senior commanders and staffs; indeed, the U.S. Army even places some specific long-range assets under his command in the event of war. The Seventh Air Force leadership, in turn, puts foremost emphasis on supporting the CINC's vision, and that attitude permeates the planning staff.

In comparison, USAFE planners suffer from a fragmented command structure. Because of both U.S. and NATO command arrangements (see Figure 3.1 for a depiction of the operational chain of command), lines of control between leaders and led can be bafflingly complex. The two primary operations ongoing in the Balkans—the *Provide Promise* aid airdrops and *Deny Flight* airspace monitoring—operate from separate ATOs; there is, in addition, a third ATO for training activity in the region. Although attempts are made to coordinate all these activities, the multiplicity of command and control paths adds unhelpful complexity—too many sheets of music inhibit true harmony, so to speak.

Additionally, European-based units and staffs are only beginning to adapt to their new role as expeditionary forces. So, limitations on the physical connectivity between people and locations impede smooth command functioning—and hence the clear articulation of intent. For example, as of fall 1993, the air component commander for NATO operations over Bosnia can talk with some of his flying units only through a communications link located at the theater CINC's headquarters, hundreds of miles from his own HQ.

Finally, measures of effectiveness for achieving objectives often lack any specific operational basis and are sometimes drawn from the individual experience of some senior commander ("Deliver *X* weapons of *Y* type on each target in category *Z*.") Although such measures of

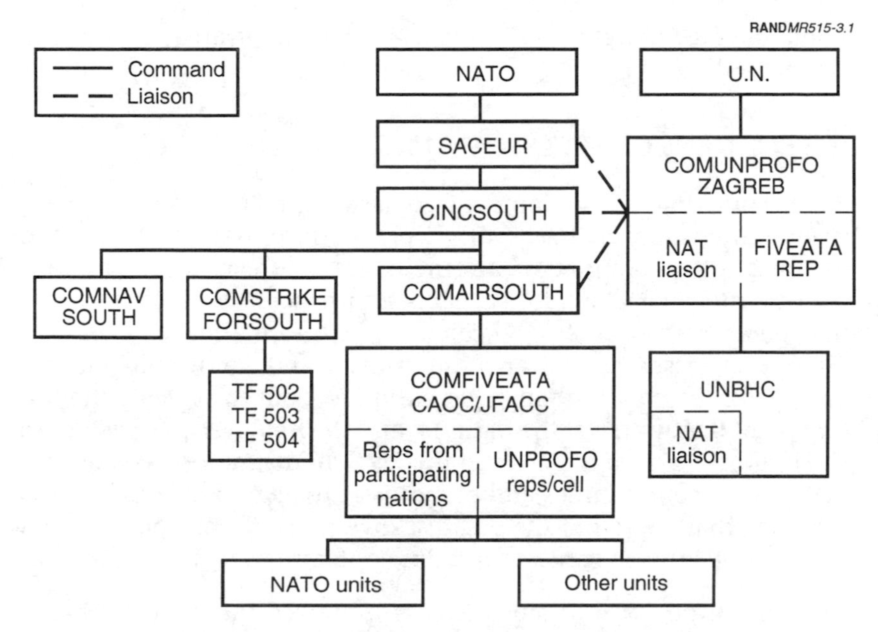

SACEUR—Supreme Allied Commander, Europe
CINCSOUTH—Commander in Chief, Southern Region
COMAIRSOUTH—Commander, Allied Air Forces, Southern Europe
COMFIVEATAF—Commander, Fifth Allied Tactical Air Force
CAOC—Combined Air Operations Center
COMSTRIKEFORSOUTH—Commander, Strike Force South
COMNAVSOUTH—Commander, Allied Navies, Southern Europe
COMUNPROFOR—Commander, U.N. Protection Force
UNBHC—U.N. Bosnia-Hercegovina Command

Adapted from "Bosnia Mission Stretches Airborne Eyes and Ears," in *International Defense Review*, January 1994, p. 55.

Figure 3.1—NATO/U.N. Operational Chain of Command for Bosnia

effectiveness (MOEs) may be perfectly valid—based as they are on a commander's expert judgment—they provide no clear basis for relating the outcome of a given attack to any effect on enemy combat capability. In many cases, mission-type orders—"Prevent advancing Iraqi units from reinforcing enemy forces in and around Khafji"—are probably preferable. Further, there appears to be fairly universal support for enhancing capabilities for what might be called functional damage assessment, an idea we discuss at greater length in Chapter Four.

INTELLIGENCE SUPPORT AND INTEGRATION

Establishing and maintaining an adequate intelligence support infrastructure is a prerequisite to selecting appropriate strategies and defining and tracking attendant measures of effectiveness and outcome. Intelligence support collects and processes information on the enemy, assesses the significance of the information, and disseminates these assessments to decisionmakers and planners in a timely and useful manner. Although we did not set out to evaluate the operation of this system at the national and theater levels, we did take a few opportunities to understand the level of integration between the intelligence communities and planners at the working level. It is in this area that we found the greatest divergence in perspectives between our Pentagon contacts and our contacts in the field.

Officers on the Air Staff repeatedly emphasized the difficulties they have experienced getting what they perceive as adequate support from the intelligence community. They firmly believe that the system is "demand-pull"—with the accent on pull, as in pulling teeth. In other words, rather than being able to draw upon a regularly replenished and updated stock of information at any time, planners on the Air Staff find themselves specifically requesting each piece of information.

During a period of crisis, this can change, as in the months leading up to the 1991 Gulf War. At that time, the CHECKMATE planning cell in the Air Staff received broad and effective cooperation from the intelligence community as they developed the initial air campaign plan (*Instant Thunder*) and supported planners in Riyadh. In this instance the ad hoc arrangement was successful, but it is clearly wasteful and potentially dangerous to force planners and intelligence staff alike to climb a steep learning curve as a crisis looms or breaks. To some extent at least, the kinds of relationships that were developed between CHECKMATE and the intelligence world should be revived and institutionalized so that Pentagon planners are better prepared to face the next contingency.

Some fear that in peacetime institutional incentives drive the intelligence community's response to demands from the operators. According to this view, making intelligence information *too* readily available risks obviating, or at least dramatically reducing, the need for specialized intelligence support. We have no definitive evidence on this issue, but we believe that such concerns are somewhat misbegotten. The actual *provision* of information to other parties is a fairly small part of the intelligence community's business, which revolves around its collection, processing, and analysis functions. Hence, even if *all* processed intelligence were available instantly and at no cost to users, the bulk of the community's institutional interests would be protected. Instead, we believe that, to quote Strother Martin in the motion picture *Cool Hand Luke*, what we've got here is failure to communicate.

The relationship between the intelligence and operational communities is, in peacetime, not particularly intimate, security concerns—the "green-door" syndrome—being one major distancing factor. The Gulf War pointed up many problems arising from this separation; unfortunately, it does not appear that the need for more common frames of reference and regularized operating procedures has retained a high priority in the postwar world. It may be the case that substantial payoffs could be forthcoming from a modest, but sustained effort to break down the walls between intelligence personnel and planners and operators.[5]

In the field, integration appears better, if still not seamless. In Europe, a reorganization is bringing the intelligence and planning staffs together under a new directorate for operations at USAFE; all parties hope that this will help bridge the gap between them. Also, in both Korea and Europe, integration at the planning-cell level is good. In both places, intelligence staff work side by side with the planners, learning their needs and at times growing adaptive enough to anticipate them.

---

[5]See the discussion of Air Operations Groups in Chapter Four. The authors have some experience in investigating these cross-community disconnects, which may predispose them to see such disconnects at work in this instance. See E. L. Warner, D. J. Johnson, D. A. Shlapak, and D. E. Thaler, *Space Support to Terrestrial Operations*, Santa Monica, Calif.: RAND MR-299-AF, 1993.

At each site we visited we heard concerns expressed that, as one individual put it, the "feedback loop" between planning, execution, and assessment "is broken." That is, accurate information about what has been attacked, and to what effect, does not percolate rapidly back to either the target-nominators or the planners. This can lead to unnecessary reattacks, or neglect of a still-functioning target. Additionally, friction can arise between target-nominators and planners when the former cannot tell from assessment reports that their input is being acted upon.[6]

There are high hopes that the Contingency Theater Air Control System (TACS) Automated Planning System (CTAPS)[7] will significantly improve the integration of intelligence and operations.[8] However, many intelligence support systems—including some still under development—do not interface smoothly with CTAPS. If CTAPS is to serve as the core architecture for air planning and execution—as it is intended to do—these difficulties must be ironed out.

On the technical level, intelligence dissemination is becoming an increasingly high-volume undertaking. More and more users are calling for access to materials such as secondary imagery, transmission of which requires high-speed, large-capacity data links. In less-developed theaters and expeditionary conditions, such capabilities may be hard-pressed to keep up with the demand from intelligence and other users. We suspect that the need for high-quality, high-capacity communications will only grow with time, as more and more data become available in near-real-time from an ever-expanding variety of sources. We realize that one can never have too much communications capacity; however, it would be unfortunate indeed if expensive—and potentially quite valuable—initiatives to rapidly provide intelligence information to planners and operators foundered be-

---

[6]It is reported that in the Gulf, the nomenclature and identifying number used to tag specific targets nominated by the Army changed as the request climbed the ladder from division to Corps to ground component, then finally to the air component. This made it difficult for lower echelons to tease out which of their requests had or had not been filled from the after-action reports they received back.

[7]A good description of TACS may be found in R. J. Blunden, Jr., *Tailoring the Tactical Air Control System for Contingencies*, Maxwell AFB, Alabama: Air University, 1992.

[8]CTAPS is described at greater length later in this chapter.

cause of a scarcity of electronic pipelines through which to push the product.

Finally, the intelligence personnel we spoke with shared with the planners the desire to develop a more functionally oriented approach to damage assessment; indeed, they recognize this as one of the foremost challenges they face as a community.

RESPONSIVENESS OF THE PLANNING CYCLE[9]

Winnefeld and Johnson articulate one commonly heard complaint about the air-operations planning process in the Gulf War:

> in one important respect, the ATO was not flexible; it took forty-eight hours to build an ATO for any given flying day . . . . The ATO was particularly well suited for use against a hunkered-down enemy who had lost the initiative. But, in a rapidly changing situation, or when there were delays in bomb damage assessment, execution problems could and did occur.[10]

The process during a campaign of setting priorities, developing a time-phased plan, and building and disseminating an ATO[11] is sometimes referred to as the planning, decision, and execution (PDE) cycle. In and of itself, a PDE cycle can run from about 36 to 48 hours, depending upon whether or not certain high-level CINC decision-making is included in it. In fact, however, the PDE cycle is itself embedded in a larger, longer process.

---

[9]This discussion focuses on planning in an ATO-oriented environment. Colleague Myron Hura observes that the USAF can also plan and execute air operations using mission-type orders, in which flying unit commanders are given a set of objectives ("provide support to engaged elements of the 2d Armored Division") and broad guidance to plan and generate the needed sorties.

[10]J. A. Winnefeld and D. J. Johnson, *Joint Air Operations: Pursuit of Unity in Command and Control, 1942–1991*, Annapolis, MD: Naval Institute Press, 1993, p. 110.

[11]Seventh Air Force in Korea refers to an ITO; we will use the more generic ATO nomenclature throughout this report.

A depiction of a notional PDE cycle is shown in Figure 3.2.[12] The process starts with a candidate apportionment being briefed to the JFACC at 1100 *two days* before execution is to begin (e.g., if this is Friday's ATO being constructed, the candidate apportionment is briefed on Wednesday morning). The CINC issues his guidance and targeting strategy at 1700—this is the "official" beginning of the PDE cycle. What follows is a set of interconnected steps in which this high-level guidance is, incrementally, translated into specific targets to be struck and aircraft to do the striking.

Rather than feeling concerned that the PDE cycle was overly prolonged, we were impressed that the development of a complete, theater-wide ATO could be accomplished, day after day, in such a comparatively *short* span of time.[13] A clue to why this is so may have been provided by an officer who commented that, for the people involved, ATO production moved on a 24-hour cycle. That is, the same people come in every day and do the same job. This routinization of the job probably contributes greatly to the various planning cells' ability to accomplish their complex orchestration of the air war in a relatively brief span of time.

There appear to be three separate threads to the arguments for speeding up the PDE cycle:

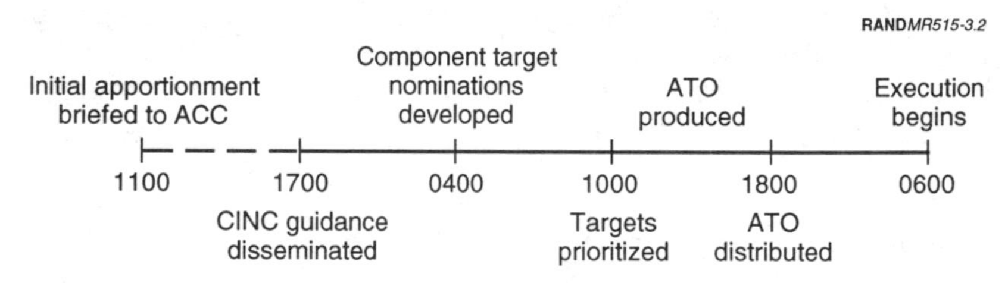

Figure 3.2—Notional PDE Cycle

---

[12]The specifics of the process, and its precise timing, will vary from theater to theater and contingency to contingency. The depiction is intended to convey a meaningful flavor of the timeline.

[13]This is especially true given the amount of retyping and "sneakernetting"—carrying of floppy disks from one computer to another—required by current procedures. In the next subsection, we will discuss initiatives that could dramatically reduce these inefficiencies.

- As Winnefeld and Johnson suggest, the cycle is too long to be properly adaptive in a rapidly changing operational-tactical situation.
- The target priorities of other component commanders get lost in the shuffle as the ATO is generated.
- The cycle does not allow enough time for wing- and squadron-level mission preparation between when the ATO is distributed and when engines start.

ATO Responsiveness

Although Iraq ceded the initiative to the Coalition in early August and made no determined effort to seize it back, the Gulf War does provide some insight into the first issue. The most notable Iraqi attempt to force the Coalition's hand—the attack in the vicinity of Al-Khafji from 29–31 January—also represents probably the sternest challenge presented to the responsiveness of allied air power. Reaction, by all accounts, was relatively swift and effective. JSTARS and AWACS aircraft provided target cueing and engagement control while Air Force, Marine, and Navy aircraft pummeled the forces in and around Khafji, the vessels carrying amphibious reinforcements, and two divisions that were moving toward contact.[14]

The battle of Al-Khafji was obviously not a particularly trying ordeal for a Coalition that by the end of January had total air supremacy and a huge pool of air resources from which to draw. However, the ATO was sufficiently flexible to accommodate the need to apply a concentration of force against an unexpected Iraqi gambit.

Indeed, we believe that *flexibility* is really at the heart of the responsiveness issue, not the number of clock-hours it takes to produce and distribute it. Logically, one tasking order could suffice for an entire campaign were it possible to craft one sufficiently plastic.[15] Con-

---

[14]See Department of Defense, *Conduct of the Persian Gulf War*, Washington, D.C., April 1992, pp. 130–133. See also R. P. Hallion, *Storm over Iraq*, Washington, D.C.: Smithsonian, 1992, p. 219–223.

[15]In Korea, the first few days of the air war are in fact preplanned so that a short- or no-warning attack does not catch the Seventh Air Force flat-footed.

versely, evolving a process that churns out ATOs in 18 or 24 hours would be counterproductive if the result were a rigid document that could not readily adapt to a rapidly evolving situation, like that in Al-Khafji.

A general sense that faster is better seemed to pervade the conversations we had on this topic; however, this opinion was not universal. Officers at one command we visited were concerned that speeding up the process too much might result in confusion as the execution and planning cycles began to converge. At another stop, planners worried that a cycle any shorter than 24 hours would tempt the JFACC and his senior staff to lose their grip on the long-term big picture and focus on day-to-day details that are properly below their level of interest. As one put it, "You can take the general out of the cockpit, but you can't take the cockpit out of the general." This perspective certainly coincides with one author's experience running and playing high-level crisis and war games throughout the U.S. military. Fundamentally, we think the assumption that "faster is better" needs to be critically examined before it becomes doctrine or, worse, dogma.

This is not to argue that current initiatives to speed up the planning cycle should be abandoned. Rather, the PDE cycle should be placed in an appropriate context. Among the questions that should be addressed are:

- Given that the PDE cycle is enmeshed with a much-longer bomb damage assessment (BDA)/target development cycle, how much time can be shaved off the former before it becomes seriously out of synch with the latter?[16]
- How much "better" can the application of air power actually get merely by being "faster"? Put differently, what are the operational objectives or tasks that we can attain or undertake much more successfully or efficiently if the ATO takes 30 hours to produce vice 37?

---

[16]We were told that the BDA process takes several days—results of attacks made on Monday typically are not fed back into the planning cells until Wednesday when Friday's operations are already being laid out. Even a dramatically shortened ATO loop could therefore not realistically support a reattack of a Monday target before Thursday.

- Finally, can flexibility be gained by taking better advantage of existing MCEs?

In short, we believe that the issue of responsiveness hinges on more than just how fast the ATO can be built. A better formulation of the challenge would embrace efforts to speed ATO construction along with revising the planning process to create a more flexible instrument for the application of air power across its full range of strategic, operational, and tactical employment. We will return to this issue in the next chapter.

Target Priorities and After-Action Reportage

In every war since the advent of air power, disputes have arisen over the proper allocation of air effort. The war against Iraq was certainly no exception. Ground force commanders—including the CINC—sought increasing control over the air campaign; as the ground war approached, demands from the ground components for more battlefield-preparation sorties grew more strident. Eventually, the deputy CINC, Army General Waller, had to be brought in to arbitrate disputes.[17]

The target-nomination process differed in detail at every command we visited, but the general drill was similar. A targeting cell at the air component command—which includes liaison officers from the ground and naval components—integrates target requests from various sources, attempting to harmonize the list with the CINC's guidance as interpreted and detailed by the JFACC. It is in this cell that Army, Navy, and Air Force disagreements can get hammered out as the theater target list for a given day is racked up. The cell also serves as one of the main interfaces between the intelligence community and the planners and pilots.

As noted earlier, feedback to the ground forces—or, more precisely, lack of feedback—is a source of some friction. In the Gulf, it was often true that ground commanders lost sight of their target nominations once the list was passed on to higher echelons. After-action

---

[17]U.S. News and World Report, *Triumph Without Victory,* New York: Random House, 1993, p. 268.

reports might denote a struck target with a name and reference number that was unrecognizable to the nominating entity, making follow-up difficult. The same disconnect could also lead a ground commander to ask for a strike on a target that had already been hit, albeit under a different label. These kinds of purely administrative difficulties amplified and exacerbated the very real struggle between components for control of the limited pool of available air resources.

Again, we suggest that peacetime practice and planning can help prevent such problems from arising in future contingencies. Each command we visited has organized or is organizing an *air operations group* (AOG), a kind of standing mini-JFACC staff. Seventh Air Force in Korea has the largest and best-developed AOG, as perhaps befits an organization that stands 24-hour watch against an imminent threat. The role that the AOG is to play in the other commands was somewhat in flux at the time of our visits. It is clear, however, that the AOG could be used as a forum for working out as many intercomponent issues as possible in advance of a crisis. Ground and naval liaison personnel should be included in the group, and procedures for managing target nomination and attack reporting should be established and practiced. We will have more to say about the value of the AOG in Chapter Four.

Wing and Squadron Mission Preparation

A final complaint about the length of the PDE cycle is that insufficient time is left for mission preparation at the wing and squadron levels. The idealized cycle shown in Figure 3.2 shows 12 hours between the time the ATO is published and when execution begins. However, several factors can in practice cut into this time. First, the ATO can be a long and complex document—it consisted of hundreds of pages in the Gulf War—and cannot be promulgated in zero time. For example, in the Gulf War, computer and other incompatibilities between the Air Force and Navy meant that the ATO had to be flown from Riyadh to the carrier battle forces in the Persian Gulf and Red Sea. Second, some aspects of mission planning must be complete well in advance of the actual beginning of the sortie so that appropriate munitions can be prepared, mission-specific software loaded, and so forth. Finally, for any number of reasons, the ATO may get completed late.

Ameliorative measures are available for all these difficulties that do not require a major reworking of the PDE process, however. Since the Gulf War, for example, the Air Force and Navy have worked to overcome their disconnects so that the ATO can be transmitted from shore to ship electronically. Distribution time might also be cut by reverting to a variant of the old "fragging" system, whereby each flying unit receives only that portion of the ATO that applies to it, instead of the entire document. Finally, an expedient employed during the Gulf War might be formalized. During that campaign, wings often sent observers to Riyadh to keep track of the day-to-day planning. These officers could tip off their units to likely taskings well in advance of the actual distribution of the ATO. This is obviously an imperfect arrangement—a last-minute change in mission could undo hours of preplanning at the wing, possibly leaving it worse off than it would have been without any advance warning. Nonetheless, combined with the other measures described, a formalized version of this informal workaround might relieve some of the perceived pressure for accelerating the PDE cycle.

AUTOMATED SUPPORT TO PLANNING

Currently, the planning process is supported by a hodgepodge of computer systems that at times seem to be at war with one another. Each planning group—airspace managers, "fraggers," intelligence staff—seems to have its own hardware and software, designed and procured with little attention to whether or not it would all work together. Among the problems highlighted to us were:

- There are incompatibilities among operating systems, with some users relying on UNIX, others MS-DOS, and still others on Macintoshes.
- Parallel difficulties exist with data media—3.5-inch diskettes versus 5.25-inch disks and so forth.
- In many cases, information must be rekeyed as it moves from station to station, which takes time and manpower.
- Information is often segregated in different systems, so that planners in the different communities—say, F-15 planners and tanker planners—cannot call up and work from a common database.

Overcoming these problems and streamlining planning demand an integrated hardware and software system that pulls together all users and all data under a common architecture. CTAPS is the main initiative in this realm.

CTAPS is a "common, powerful, computer system architecture that adheres to joint standards [consisting] largely of off-the-shelf hardware and software."[18] CTAPS is hosted on Sun Microsystems SPARC computer workstations connected on a local area network. It is designed to facilitate one-time data entry, with all information available to all users on an as-authorized basis (e.g., F-15E planners may be able to view, but not change, tanker-related information). It provides for automatic transmission of the ATO and will be interoperable with automated command and control support systems deployed by the Army, Navy, and Marines. Plans, intelligence, airspace management, and execution modules are among those included in the system.[19]

We found universal enthusiasm for CTAPS among the groups we visited. All felt that it would simplify their jobs and speed up the planning cycle (albeit by some estimates only marginally, by shaving off two to four hours). Planners in one command were concerned that system designers had not solicited sufficient input from those who would ultimately use CTAPS, resulting in some design decisions that made more technical sense than they did operational; when asked about this, however, respondents at other commands disagreed with this assessment. Seventh Air Force staff expressed an ongoing concern that the system be configured so that it could be fully exploited in their environment, which involves close contact and interaction with allied personnel; to them, it was imperative that CTAPS be "releasable." They also expressed a desire for "CTAPS in a box," a highly portable, easy-to-set-up system that could be employed at the many collocated operating bases (COBs) that would receive U.S. tactical aircraft in the event of war in Korea.

---

[18]*JFACC Primer*, op. cit., p. 36.

[19]As colleague Myron Hura notes, CTAPS is a group of systems, many of which are still being developed; many challenges remain. For example, the heart of the CTAPS initiative is integration of diverse air-planning activities, but many of the software and hardware subsystems needed to achieve a high degree of integration—such as secure, high-speed, local-area networks and imagery-dissemination systems—are underfunded.

An initiative that shows promise in the area of objective definition and articulation is the air campaign planning tool (ACPT) being developed in the Air Staff. The ACPT is a minicomputer-based system that draws the user through the planning process using an objective-based hierarchy similar to the structure discussed at the beginning of Chapter Two. The planner works through the hierarchy level by level, beginning with national military objectives, and wending downward toward specific targets. The tool helps the user link objectives to one another and also highlights areas where connections, or the objectives themselves, are missing or unclear. The user also sets priorities among objectives, and draws on a large database to identify appropriate targets for achieving specific goals. The end result is an overall prioritized target list that is linked up through the hierarchy to overall campaign goals.

It is hoped that the ACPT will help institutionalize the top-down logic and keep commanders "out of the weeds." The ACPT is intended to be fully compatible with CTAPS and exists currently in a prototype form; we believe that, properly evolved, it could make a great contribution toward ensuring that means and ends are properly connected in future air campaigns. Someday, it may aid planners in conducting quick-turnaround "what-if" analyses during campaigns to look a week or so into the future.

It was emphasized by a number of planners that one should not go overboard in one's fondness for or hopes about automation. As argued earlier, trading flexibility in planning and execution—a characteristically human trait—for electronic efficiency would be a bad bargain. Planning will remain a manpower-intensive process, much of which—setting and decomposing objectives, identifying important excursions, flagging events or trends that are, for good or ill, "out of bounds"—will continue to require smart, well-trained people with grease pencils or white boards. The goal of automated support to planners should be exactly what the term itself implies—providing ready access to information and easier recording and implementation of decisions. It does not appear that the time or technology are yet ripe for the machine to supplant—as opposed to support—the man in the planning loop.[20]

ORGANIZATIONAL PERSPECTIVES

We found the four principal organizations we visited—Ninth Air Force, USAFE, Seventh Air Force, and HQ PACAF—to be in substantive agreement on most of the issues we discussed—intelligence support, prospects for automation, and so forth. We did, however, find some differences in tone and approach that are worth noting.

At Ninth Air Force, we found an organization that had recently fought and won a major conflict. Commendably, planners there seemed focused on learning the lessons of the Gulf War. Other than welcoming CTAPS, however, they seemed less interested in other major changes to planning processes or procedures than were staff in Europe and Korea. Their experience in the Gulf seems to have inspired confidence that they can successfully plan and execute a campaign when called upon.[21]

A unique emphasis we found at Ninth Air Force was on the power of personalities to affect a campaign. This perspective, again, appears strongly informed by the organization's experience in *Desert Shield* and *Desert Storm,* when new procedures, organizational arrangements, and personnel were overlaid from the top down on existing CENTAF structures. Although the ad hoc arrangement worked—in the sense that the air campaign was planned and executed with great success—the price included what some perceived as

---

[20]Alan Vick observes that industrial and commercial experience shows that "automation only makes sense if the process being automated is logical, coherent, and streamlined to begin with." Deriving maximum benefit from CTAPS and related initiatives requires that the planning system—and its goals—be properly defined and well understood. "Paving the cow-paths" by automating procedures that are outmoded or ill conceived brings little advantage.

[21]The role of the Air Staff, and particularly Col. John Warden's CHECKMATE branch, should not be neglected in discussing Gulf War air campaign planning. Early on during *Desert Shield,* Ninth Air Force was extremely busy managing deployment and beddown issues, and CHECKMATE's *Instant Thunder* concept helped fill something of a void in warfighting planning. Later, after CENTAF was settled into Riyadh, it continued to receive support from CHECKMATE and other Air Staff organizations throughout *Desert Shield* and *Desert Storm.*

needless friction between individuals and organizations. The lesson here seemed to be that the best laid plans, structures, and associated training can be easily overturned by commanders who, rightly or wrongly, have strongly held convictions about what is needed to support them. Planners everywhere must be prepared for such situations.

Seventh Air Force's perspective was shaped by the history and geography that places the command in the center of a well-developed theater, with an up-and-running coalition and a viable threat literally minutes away. Of the three commands, they most strongly emphasized the importance of joint and combined operations. They seem wholly committed to supporting the CINC's plan, even to the point of decrying the use of the term "air campaign"; only the theater commander has a campaign plan, according to several officers we spoke with. Components plan *operations.*

Planners in Korea benefit from a stable and ongoing relationship with their South Korean counterparts. As we toured various facilities at Osan AB, we found U.S. and Korean personnel working side by side virtually everywhere. While such proximity may create some peacetime administrative headaches—difficulties dealing with certain kinds of sensitive intelligence, for example—it should mean a faster and smoother spin-up in the event of a crisis or conflict on the peninsula.

While the North Korean air force may represent a minimal threat to USAF operations, the North's special forces (SF) may pose a significant danger. As the Air Force manpower pool shrinks, security police forces are being likewise reduced; much responsibility for air base security will rest with Korean reserve units or reinforcing U.S. troops, and neither one may be available in time to fend off an initial North Korean SF onslaught. Planning should take into account the desirability of engaging and neutralizing enemy SF in transit or while they are assembling rather than waiting to deal with them on and around their target air bases.

Friendly air operations could also be affected by "sleeper" agents—South Korean civilians who, in the event of war, would carry out various acts of sabotage and espionage in suppport of the North's military offensive. Similarly, air bases could become magnets for floods of refugees, both Korean and U.S., seeking sanctuary from impending or just-begun hostilities. Planners should take these challenges into account and prepare in advance for potential disruptions in base operations originating from them.

Base defenses in Korea are tested in the annual *Foal Eagle* exercise, and innovative concepts, such as owner-user responsibility for point defense, have been implemented. Despite this, several officers observed that, judging by their experience both in Korea and elsewhere, the Air Force as a whole seems disinterested in the problem of air base security. In peacetime, security is perceived as an expensive, manpower-intensive nuisance; in the event of war, however, the successful conduct of sustained air operations could be jeopardized if the north's "second front" is even modestly successful. If, as some suggested, the initial battle for air supremacy over the Korean battlefield will be won or lost on and around allied air bases, some careful attention should be paid to ensuring the protection of those facilities. In fact, we believe that air base security should be given higher priority worldwide. Future adversaries who cannot hope to meet U.S. air power head on will certainly look to attacks on its supporting infrastructure in an effort to counter it.

That said, we are concerned that planning for the initial period of the air campaign in Korea account more for the possible effects of air base *in*security. In particular, we hope that plans for Korea anticipate the possibility of reduced sortie rates, the commitment of a large number of ground-attack missions to rear-area targets, and the operational effect of shifting command and execution responsibilities to alternative locations in the event Osan is cut off or otherwise incapacitated.

We are concerned that current deployment plans may be somewhat inflexible and therefore vulnerable to derailment by the kinds of disruptions in the allied rear area that we have described. This danger could be especially severe in the event of a short-warning attack, already a very stressful scenario. Exercises should take these possibilities into account so that any deficiencies can be identified and addressed before a crisis flares.

Finally, we turn to Europe, where a great transformation is in progress. With the enormous changes in the European security context, USAFE finds its setting and responsibilities in flux. For forty-plus years, U.S. forces in Europe planned against the Warsaw Pact; their circumstances then were in many ways analogous to those prevailing today in Korea—imminent threat, in-place coalition, well-understood planning concepts and objectives, and a "fight-in-place" mind-set. Today, their situation bears more than a passing resemblance to that of the Ninth Air Force; USAFE is now the air component of an expeditionary force that must prepare to fight anywhere, anytime, with a pickup team of allies. This is a fundamental, and wrenching, kind of change, and USAFE is faced with muddling through as best it can.

Further complicating matters is a whole cluster of issues about NATO. Like USAFE, the Alliance is in transition, and there is much confusion about its role in the "new" Europe. On a political level, NATO's structure and mechanisms may be ill suited for dealing with the kinds of issues the Alliance is now confronting. For planners, this means that command authority may be diffuse and confusing; adding the U.N. to the mix—as has happened in the Balkans—adds an additional layer of complexity.

Among the many challenges facing USAFE today is a lack of clarity about objectives. Planning for today's limited wars, peacekeeping operations, and so forth is conducted in something of a vacuum compared to the solid framework of enduring objectives that characterized the Cold War. When U.S., NATO, and U.N. leaders are unable to articulate coherent goals for military operations, or those objectives appear fluid and transitory, planning becomes difficult, to say the least. In such circumstances, planners can be forgiven if they sometimes see themselves as the sculptor's apprentice who, having been told to bring his master "a rock, any rock," drags one back to the studio only to be told that it is the "wrong" rock.

USAFE is obviously not the only organization affected by the evolution of U.S. foreign policy into its eventual post-Cold-War form, and planners there—and everywhere—must be adaptive to the changed world by responding creatively to the circumstances they confront. Nevertheless, we wish to remind readers that no doctrinal changes, improvements in automated data-processing, or quantum leaps in intelligence support can fully substitute for well-defined, clearly stated objectives as a basis for sound and effective campaign planning.

Chapter Four

REFINING THE AIR CAMPAIGN PLANNING PROCESS

In this chapter, we focus on four challenges for future analysis derived from the observations in Chapter Three.[1] These four challenges are to

- properly define, prioritize, and determine the relevance of theater military objectives in a variety of scenarios
- enhance intelligence support to commanders, planners, and operators
- improve the responsiveness and flexibility of the campaign execution cycle (from adjustment of strategy to development and dissemination of the ATO to mission execution)
- gain maximum advantage from the AOG.

OBJECTIVES: DEFINITION, PRIORITIZATION, AND RELEVANCE

In the preceding chapter we noted that planners sometimes must ply their trade in the absence of clear guidance from higher echelons.

---

[1]RAND is doing extensive work in several of the areas under discussion. See, for example, Myron Hura and Gary McLeod, *Route Planning Issues for Low Observable Aircraft and Cruise Missiles: Implications for the Intelligence Community,* Santa Monica, Calif.: RAND, MR-187-AF, 1993. See also Myron Hura and Gary McLeod, *Intelligence Support and Mission Planning for Autonomous Precision-Guided Weapons: Implications for Intelligence Support Plan Development,* Santa Monica, Calif.: RAND, MR-230-AF, 1993. There are also ongoing studies of dynamic battlefield management and surveillance and targeting of critical targets.

We believe that, in a number of circumstances, this will be the norm rather than the exception. Often, planners will find themselves suggesting objectives to higher authority—i.e., to those who ideally should define those objectives. Accepting the likelihood of such circumstances, the issue becomes one of helping planners better do their jobs in such an environment.

A crucial challenge, then, is to provide remedies for situations where national- and theater-level objectives are not well defined or where cause-effect relationships between military options and desired political results are unclear. This points to the need to build a menu of potential campaign and operational objectives in various scenarios, to gain insights into appropriate priorities among these objectives, and to link the achievement of these objectives to political aims.

The first task is probably the least difficult, particularly for scenarios involving the defense of friends and allies against large-scale conventional attacks. Iranian or Iraqi aggression against Saudi Arabia or a North Korean invasion of the south are obvious examples. A menu of campaign objectives for defeating the aggression in these scenarios might look as follows:

- Gain and maintain air superiority or supremacy
- Halt invading armies
- Deny or counter enemy use of weapons of mass destruction
- Gain and maintain sea control
- Suppress enemy war-supporting infrastructure.

Building menus for "lesser" contingencies such as Bosnia would likely prove more challenging, as would the analyses these menus would drive. But with the potential for more Bosnias and Somalias on the horizon, as well as the increased emphasis on peacekeeping and humanitarian missions, such lists of objectives should be quite helpful to planners of air and other operations.

Drafting menus of objectives for exemplar scenarios would provide a baseline against which to plan air campaigns, especially at times when guidance from above is insufficient. We term this a baseline, of course, because planners would tailor the menus to specific contingencies. For the menus to be useful in this regard, then, they would need to cover large areas of the potential scenario space.

Possessing well-scrubbed lists or menus of objectives is by no means enough. The menus should also serve as a focus for campaign analysis. Such analysis would provide insights into the two related issues mentioned above, namely: What is the appropriate weight of effort to be applied to operational and campaign objectives over time, and how does the achievement of these objectives help secure political aims? The first issue has been addressed quite extensively at RAND and elsewhere with regard to major regional contingencies; much less has been accomplished dealing with other types of scenarios. The second issue—e.g., how would silencing the Serb guns in the hills around Sarajevo help encourage the Serbs and Muslims toward a negotiated settlement?—is far more challenging yet is critical to the future application of U.S. military power.

It seems clear that the quantification of alternative weights of effort among objectives in different scenarios would help planners a great deal. Such quantification would enable planners to run "what-if" analyses at the operational and campaign levels, and may be incorporated into such emerging automated aids as the Air Campaign Planning Tool. Figure 4.1 represents the kind of analytic results that we believe would be helpful in this regard.

Assume three campaign objectives—A, B, and C—and a desired political aim. The ordinate (Y-axis) of Figure 4.1 shows the number of days to attain the political aim, the abscissa (X-axis) gives the weight of effort applied toward objective A, and three weights of effort are shown for objective B. The remaining weight of effort is applied to objective C (e.g., at 25 percent for A and 10 percent for B, or 35 percent, the weight of effort applied toward C is 65 percent).

To draw such relationships, one must first understand the most salient military and political objectives, links between them, and the appropriate measures of merit. We believe that a concentrated analytic effort along the lines described above would greatly improve the U.S. ability to plan air operations in support of theater campaigns.

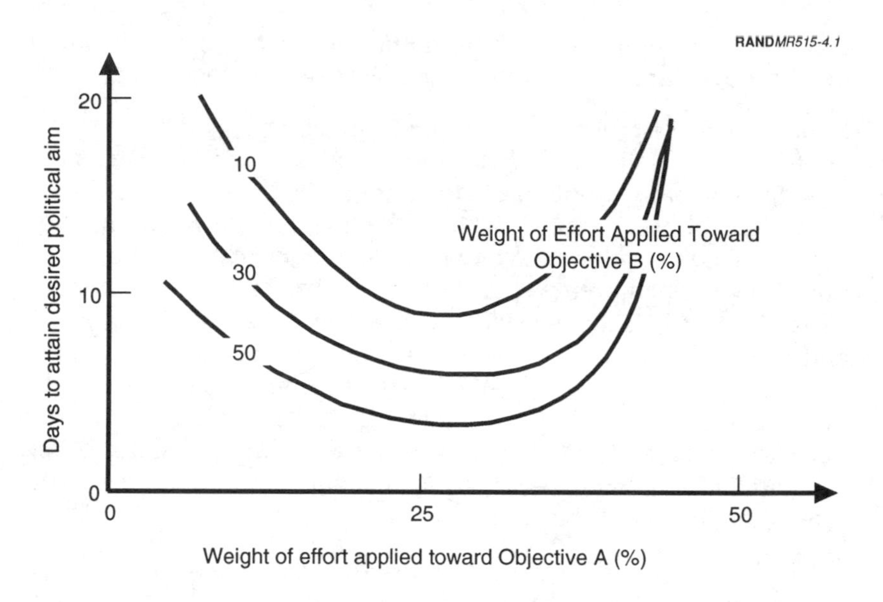

Figure 4.1—Exemplar Results of Campaign Analysis

A FUNCTIONAL FOCUS ON ENHANCING INTELLIGENCE SUPPORT

In many ways, the treatment of objectives above is closely linked with efforts to meet the second challenge—enhancing intelligence support to commanders, planners, and operators. To gain insights into appropriate weights of effort among objectives, analysts must develop yardsticks for measuring the achievement of operational and campaign objectives. However, these yardsticks would be useless to commanders if the "vernacular" of the reports they receive on battle results and that of the yardsticks according to which they measure achievement of objectives were incongruous. Obviously, this places a burden on the intelligence community to formulate new methods for interpreting and reporting the effects of friendly and enemy action.

Moreover, many planners we encountered spoke of the need for a greater focus on *functional* as opposed to *physical* results of battle. Commanders and planners need to know the effect of their actions on enemy capabilities, not merely how many items of enemy equipment are "confirmed kills." They require information about the status of a target system, how the status is changing, and how this relates to attainment of the commander's objectives. Only then can commanders adjust their strategy in the most effective way.

We believe that functional battle assessments should be attuned to each rung of the hierarchy of objectives described in Chapter Two. At the lowest level of operational tasks, for instance, functional assessments may help reduce the number of unnecessary re-attacks on the same target. For example, let us say that the task is to render a bridge impassable. If a *physical* assessment of the results of an initial attack reveal that the bridge remains standing, a commander might order re-attacks until his forces drop the bridge. Alternatively, a *functional* assessment might reveal that, since the initial attack, enemy forces had been approaching the bridge only to turn around without crossing. This is a sign that the bridge is impassable even though there appear to be no visible signs that the attack has compromised its structure.

Fundamentally, intelligence analysis should focus on developing new concepts for assessing the output of a targeted entity or system, not its physical integrity. As an example, let us go to the next higher level of the hierarchy, that of operational objectives. One important operational objective is to suppress enemy surface-to-air defenses. A focus on the physical attributes of an air defense system might lead one to count the number of surface-to-air missile (SAM) sites or radar stations destroyed. However, the output of an air defense system lies in the number of radars illuminating, SAMs fired, and friendly aircraft downed. If the destruction of a handful of sites induces the remaining operators to shut down their radars for fear of attracting an attack, these "outputs" will drop dramatically. Yet a damage-assessment approach that focuses only on the number of targets destroyed would miss this critical effect.

Given the importance many planners attached to enhancing the type of intelligence support provided to them, we contend that a focused research effort should be undertaken with the goal of developing new concepts for functional battle assessment. This effort should be in synch with the effort to define, prioritize, and determine the relevance of objectives for various scenarios.

MAKING AIR CAMPAIGN PLANNING MORE RESPONSIVE

In the previous chapter we touched on both the apparent consensus that the air planning process—specifically, the ATO development cycle—needs to be speeded up and some of our concerns about the issue. Here, we would like to suggest a concept for an alternative approach.

First, at the risk of repetitiveness, let us be clear about the objective: It is *not* just to prepare the ATO more quickly but to develop a planning process that (1) provides timely enough outputs to allow those charged with generating sorties to do so, (2) allows the CINC and component commander to have oversight of the overall campaign architecture, (3) provides greater visibility into both the planning process and execution outcomes to the other components, and (4) is sufficiently adaptive to permit appropriate responses to changing circumstances in the battle space.

Figure 4.2 portrays a functional breakdown of one potential approach that fits this model. The *air campaign plan* is formulated by the JFACC in accordance with guidance from the CINC. It is modified only rarely, and specifies:

- Campaign phasing (according to the CINC's intent)
- Operational objectives
- Weight of emphasis across operational objectives over time
- Measures of outcome for assessing the accomplishment of operational objectives.

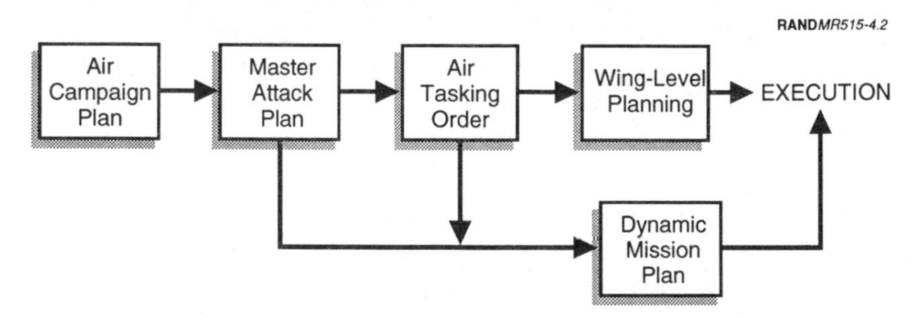

Figure 4.2—Notional Air Operations Planning Flow

The *master attack plan* is used by the JFACC and his director of combat plans to provide more specific guidance to the ATO builders. It is thus similar in purpose to the Air Guidance Letter used by General Glosson to direct daily ATO construction during *Desert Storm* but differs in that it is intended to be modified only on as as-needed basis and hence is less specific about tactical-level details (how many of what kind of aircraft will patrol which combat air patrol stations and so forth). In other words, it would not specify platforms, weapons, and targets. Instead, it represents the JFACC's operational vision of the campaign (just as the air campaign plan expresses his strategic interpretation of the CINC's concept), focusing on:

- Operational tasks to be accomplished (e.g., destroy aircraft in hardened shelters and destroy air base fuel stocks) to achieve each operational objective (in this case, suppress the generation of enemy air sorties)
- Weight of emphasis across tasks over time (emphasize attacks on air defense command posts initially, shifting to air base attacks as the surface-to-air environment becomes more permissive)
- Allocation of air sorties among tasks
- MOEs for assessing task accomplishment.

The ATO remains the centerpiece of the process. Its primary focus is on generating the sorties needed to support the JFACC's emphases and allocations. So, it would remain the tool for airspace management, tanker operations, and so forth. Also, it would translate the MAP's allocation and apportionments into specific sortie requirements for each flying unit, providing targets for those going against fixed aimpoints and tasking all others to a particular *mission control element*—such as an AWACS, ASOC, or CRC—rather than a target.

Working off of the JFACC's prioritized task list—embodied in the MAP—MCEs would orchestrate these sorties according to a dynamic mission plan, which is outlined in advance (based upon the MAP and adjusted according to the ATO) but reworked in real time by each element for its area and sphere of responsibility. For example, an airborne command and control center mission controller responsible for coordinating close-support in a given sector would know, in advance, (1) the JFACC's air-to-ground priorities (e.g., advancing armor has first priority, moving logistics vehicles second, and so forth) from the MAP, and (2) the number of sorties under his control over time, from the ATO. The controller would work out an initial plan for applying resources against the priorities; this structure would define the basic structure within which he would improvise as the day progressed. Where appropriate, this dynamic planning process would be joint, with the Army's battle coordination element and Navy and Marine liaison officers playing fully in it. Sorties could be diverted to or among MCEs in response to events.

Figure 4.3 shows a simple example of how this might work out. An F-117 strike against the Iraqi Air Force headquarters in Baghdad would be allocated and targeted in the ATO. An A-10 flight would be allocated in the ATO to report to a specific MCE—in this case, an ASOC—at a certain time. The MCE would then dispatch the flight to a target in accordance with the JFACC's priorities and the ongoing flow of events. Our final example is a two-ship flight of F-15Es "Scud-hunting" in the western desert. They, too, would be allocated to a given MCE at a given time—perhaps a controller in a JSTARS. He would then send them off against a target in the same way his counterpart in the ASOC tasked the A-10s.

Wing-level planning would look very much the same as it currently does, although increased automation should both reduce the time needed for air crew preparation and increase the amount of timely information—about weather, target status, threats, and so on—available to the crew and the aircraft/weapons systems.[2] Ultimately, crews may be able to "pre-fly" their sorties using simulators fed up-to-the-minute information by the same automated dissemination systems employed in mission execution.

Although this concept requires much exploration and analysis, it may help to increase the flexibility and responsiveness of air power in the battle space of the future by decentralizing the tasking of specific platforms. It exploits recent revolutionary advances in

---

[2]As both platforms and weapons systems (including jammers and other self-protection devices) become more sophisticated and "intelligent," building "smart," high-speed interfaces between them and their data sources will become increasingly important.

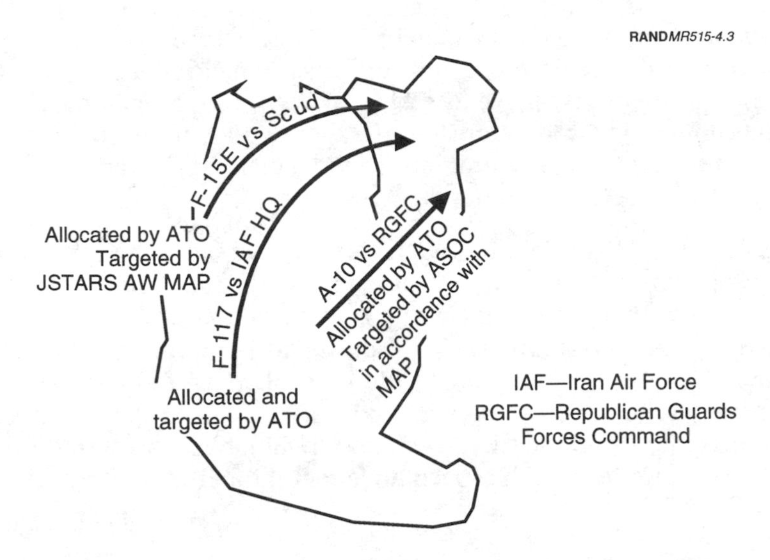

Figure 4.3—Division of Responsibilities for Allocation and Targeting

surveillance and data-processing technology whereby intelligence information can flow rapidly to dispersed customers and bookkeeping on the flow of battle can remain centralized. Massive amounts of computing power and data storage are now available in compact systems that can easily be accommodated in airborne command and control platforms. Similarly, real-time data collection—from JSTARS and U-2 aircraft, satellites, and unmanned aerial vehicles, among other sources—is rapidly coming of age. We suggest combining applications of these technologies into a system for managing responses to the fluid ebb and flow of battle. This concept seems to address the central responsiveness challenge arising from post-mortems of the Gulf War experience and, we believe, merits some analytic attention.

EXPLOITING THE AOG

The air operations group can serve at least four important functions in preparing the ground for more effective air campaign planning. As noted in Chapter Three, in its guise as an embryonic JFACC staff, the AOG can serve as a vehicle for resolving disputes and disconnects between communities (intelligence and operations) and components (air and ground) before an actual crisis or conflict brings the issues to the fore in a more dramatic and costly fashion. Properly constituted, the AOG might also serve as a kind of institutional memory of how such difficulties were addressed, so that as personnel rotate in and out, the corpus of the problem-solving experience remains intact and accessible to the command as a whole.

Second, the AOG might be given a long-range planning role, providing the command with a cell that is looking ahead, beyond day-to-day issues. In this guise, the group might evaluate a range of possible scenarios and craft menus of objectives and potential responses for each. The AOG could conduct the kind of what-if assessments described earlier in this chapter, and could provide the component commander—and the CINC—with an invaluable head start if events move a scenario from the speculative realm into the "real world."

Third, building on both of the above, the AOG will fill an obvious training role for JFACC staff. The group should be fully outfitted with the same tools—the ACPT, CTAPS, and so forth—that would be used in a wartime environment. Frequent exercises—when possible involving joint (and, where appropriate, allied) participation—could be undertaken, perhaps in conjunction with the group's long-range planning activities.

Finally, the group could be used as a laboratory for testing new planning and execution concepts. Again, building on its potential as a forward-looking planning element and a training center, the AOG could "play-test" innovations. For example, the AOG in Europe could—possibly in conjunction with the Warrior Preparation training facility—evaluate the practicality and payoffs of the scheme for dynamic mission planning described above. The innovations would of course be well scrubbed ahead of time so as not to overburden the AOG with testing numerous "half-baked" concepts. We suggest that the possibility of using the AOG in this manner be examined.[3]

Clearly, the AOG has a great deal of potential to fulfill a variety of roles for the air component. Needless to say, however, the tempta-

---

[3]USAFE activated the 32d Air Operations Group on 19 August 1994. Part of the group's tasking seems to be in line with this recommendation. See "AOG provides quick reaction," *AFNS Review*, 29 August 1994, p. 5.

tion to overtax it must be resisted. We suggest that group staff be isolated as much as possible from the day-to-day staff routine so that they can focus their energies on their key—and unique—responsibilities. Also, each group should develop a routine, or curriculum, suited to their own particular situation and, again to the extent feasible, stick to it. Putting each staff member through a cycle of train-plan-exercise/experiment twice during his tour—once as a student, then as a teacher—might be a goal.

Chapter Five

CONCLUDING REMARKS

In sum, our conversations with USAF planners around the world lead us to believe that the system is not badly broken. There *are* problems. Bits of the planning process—particularly the complex set of linkages between objectives, target selection, and damage evaluation—remain troublesome and merit serious attention. And, as we have mentioned several times, the emphasis we saw on speeding the PDE process should be re-evaluated in a context wherein the specification of the problem does not foreordain the solution. That is, a different perspective on what is needed—more flexibility and responsiveness to rapidly changing situations, rather than a faster cycle per se—may lead to the identification of new and innovative answers.

The ongoing technological revolution offers opportunities to dramatically change the way air planners do business. If properly developed, systems like CTAPS can integrate a variety of heretofore disparate functions into an architecture that provides planners with a degree of support previously unknown. Other tools, such as the ACPT, will help future air commanders and their staffs choose between alternative courses of action with greater awareness of and connectivity to overall national and theater goals and objectives. Finally, organizational concepts like the AOG may allow air components to get a head start on planning for possible future contingencies, overcome institutional divisions, and experiment with new ways of planning and executing air operations.

We suggest that one useful near-term research agenda in this area could focus on the following four general issues:

- Defining, prioritizing, and establishing the political relevance of military objectives for a wide range of scenarios.
- Developing new concepts for functional assessment of the results of battle.
- Developing new concepts for improving the flexibility of the planning, decision, and execution process.
- Identifying and refining options for full utilization of AOGs.

Air planners have not rested on their laurels from the Gulf War, nor should they. A rapidly changing world offers numerous challenges, as well as opportunities for devising new ways of dealing with them. Both the challenges and the opportunities must be embraced if air power is to continue to fulfill its role as a key guarantor of national security.